How
is

I0142229

God
Involved
in
Evolution?

Joe P. Provenzano

Ron D. Morgan

Dan R. Provenzano

En Route Books and Media, LLC

Saint Louis, MO

⊕*ENROUTE*
Make the time

En Route Books and Media, LLC

5705 Rhodes Avenue

St. Louis, MO 63109

Contact us at

contact@enroutebooksandmedia.com

Cover Credit: Sebastian Mahfood

ISBN: 979-8-88870-054-9

Library of Congress Control Number:
2023940619

Acknowledgments

There were so many people with whom we have discussed these questions and ideas over the past 50 years that we could never list all of them, but we thank all of them. Of course, we can and do thank our wives: Linda, Sandy, and Traci for their time, support, understanding, and love. Linda also served as a non-technical reader helping us make each draft revision more understandable for non-technical readers.

Table of Contents

Preface

How is God involved in evolution? It sounds like a fairly simple question, but trying to find an answer to our satisfaction led us into what became a 50-year journey. This book is a short, non-technical summary of that journey.

We explore the currently available explanations about how God is involved in evolution and explain why a new approach is needed. Along the journey, you'll see that we made a few unique connections between some existing key ideas, got a couple of new ideas, worked together, survived a "blowout" argument, and went through many twists and turns.

In the end, you will understand why our explanation provides a reasonable answer to this question. We, like many others, are suspicious of anyone claiming to have "the answer" to mysteries like this. By "a reasonable answer," we mean an explanation or insight that leads to new insights and new connections within and among disciplines. You'll see how our explanation is consistent with traditional religion and modern science, and why having a reasonable answer to this question is so important today.

For those of you who want to follow-up with more of the technical details of the science and theology supporting these ideas, we recommend our previous book:

The Fallen Angel Model:
Deeper into the Mysteries

published in 2021 by En Route Books and Media, LLC.

Meet the Authors

The Philosopher — Joe P. Provenzano has an M.S. in Physics and is the author of *Conscious Energy and the Evolution of Philosophy* (originally published as *The Philosophy of Conscious Energy: Answers to the Ultimate Questions)* and *How to Believe in God and Science: In Three Easy Steps.*

The Deacon — Ron D. Morgan is a deacon in the Catholic Church and is a lifelong follower and student of physics.

The Scientist — Dan R. Provenzano has a Ph.D. in Applied Physics from the California Institute of Technology (Caltech). Dan works with lasers and fiber optic sensors.

Joe, Ron, and Dan are also co-authors of *The Fallen Angel Model: Deeper into the Mysteries.*

Joe and Ron are retired and live near Dallas, Texas. We have been brothers-in-law for over fifty years and have been developing and refining the ideas and insights in this book for most of that time.

Dan is the son of Joe and his wife, Linda. He is currently an Optical Scientist working in Blacksburg, Virginia. Dan is married with three children and has been involved with these ideas all his life. He wrote the *ProWave Interpretation of Quantum Mechanics* when he was in graduate school in the late 1990's.

Chapter 1

Young and Curious (1968)

In the late 1960's Joe and Ron were in their early 20's and happily married to two sisters: Joe to Linda and Ron to Sandy. Joe and Ron were both Christians and believed in the basic teachings of traditional Christianity (and still do). The brothers-in-law also had a unique combination of interests and ways of thinking. Joe was studying Western philosophy: Greek, Middle Age, and European, whereas Ron was more interested in Eastern philosophers and their approach to finding truth. Joe had a physics and math background and was very logical. Ron was more artistic and mystical but also studied chemistry and was very interested in physics.

Ron: I remember a party at Joe's parents' house when we first met. I loved going over to the Provenzano's because I knew there would be delicious food including wild game, ducks, geese, moose, and everything would be cooked with liquor. Before dinner that evening I was talking to Joe and knew he was studying physics at LSU, so I asked him to help me understand something from vector analysis. I didn't understand what a gradient was. Joe tried to explain it to me by waving his hands in a flowing manner like the current in a river. It didn't really help. I kept studying on my own and made very little progress. I had a passion for physics, but not the gears to get it. I did something strange. I bought: *An Introduction to Quantum Field Theory,* by Peskin and Schroder, which is a graduate physics text covering relativistic quantum mechanics, quantum electrodynamics, and Feynman diagrams. It cost about a hundred dollars which was a lot of money in those days. What the hell was I thinking? I couldn't understand a single word in that book. Actually, I bought a lot of other books like that one, which were expensive university texts. I still find it weird that I was driven to do this. Especially since Joe was the one who had the master's degree in physics.

We did some fun things together, but there was always this strong drive to get fundamental answers to the hardest questions. In later years we both agreed that we had never met anyone like each other. We both had a determined drive to try to understand the basic nature of reality, the creation and evolution of the universe, and all the "unanswerable" questions. We have always been like that and continue to be like that now. How we managed to marry two sisters and meet each other is one of the many great "coincidences" in our lives.

Joe believed that there was one area in traditional Christianity that was not consistent with what we know from evolution. We'll cover this area in detail in several of the remaining chapters. Ron was extremely curious about the guiding principles of evolution which led to increasing complexity. We both believed in Christian doctrine and didn't want to change it, but we realized that something was missing or maybe not worded in the best way. We understand "doctrine" in the traditional way, i.e., the way Christian beliefs are worded and taught. These beliefs predate science and sometimes are worded in ways that seem to, but do not necessarily, contradict science. This issue will be covered in more detail later. We strongly believed that God created the universe, but we wanted to understand *how* He could have done it in a way that is consistent with what we know from modern science and evolution. So, we started to take a closer look at exactly what modern science was telling us.

We learned that our space-time universe began with no matter and even the laws of physics weren't fully formed yet. From that initial state, everything we see around us evolved over a period of some 13.8 billion years. There are several key scientific findings that we need to mention here.

About a hundred years ago, scientists discovered that the fundamental "particles" of nature, like electrons and photons, move in ways that don't appear to be physical or logical. These scientists developed the theory of quantum mechanics (QM) that can successfully predict where moving particles are likely to be measured but does not describe their behavior before a measurement is taken. In order to develop this theory, they had to treat the particles as if they had wave-like properties that somehow guide their motion. This issue is referred to as the "wave-particle duality." All this begs the question: What's really going on when fundamental particles are moving? There is no universally accepted answer to this question nor is there any physical description of their motion. In fact, many, if not most, physicists say that we can't ask that question, all we can do is predict the physical outcomes. We didn't like that response because we wanted to understand the underlying reality.

Stephen Hawking described another key finding of science, called "fine-tuning," very well, when he said: "The laws of science, as we know them at present, contain many fundamental numbers, like the size of the electric charge of the electron and the ratio of the masses of the proton and the electron …. The remarkable fact is that the values of these numbers seem to have been finely adjusted to make possible the development of life." — *A Brief History of Time.* Sometimes fine-tuning is referred to as "coincidences."

Many interpret fine-tuning as an indication of a designer. Others have argued that there is no design and resort to claiming that there must be an infinite number of undetectable universes, with all possible values of all the constants. That is a philosophical argument, not a scientific argument because it can't be verified by experiment, and it has never made any sense to us. On the other hand, we found the fine-tuning argument to be a very powerful argument that our universe shows clear scientific evidence of design. We provide a technical discussion of the fine-tuning argument in our book *The Fallen Angel Model*. This is an example of how that book provides technical support for the issues briefly covered in this book.

In the next chapter we cover the various ways that people have approached the question: How is God involved in evolution?

Chapter 2

On the Table (1970)

One of the first things we noticed is that before modern science and evolution were developed there weren't any significant issues with Christianity and the science of the time. However, that changed when modern science and evolution challenged some Christian doctrines. The era of modern science began when researchers developed instruments that could make precise measurements in the physical world and then use the results to develop mathematical models and theories about the physical world. The predictions from these theories were so successful that many claimed that God was no longer needed to explain anything. Christians responded and did resolve most of the issues to their satisfaction, most but not all. So, the challenge has become how to make sure all the beliefs of Christianity can be worded in a way that preserves the basic principles of Christianity but does not contradict what we know from science.

One of the most threatening challenges was the claim that since we evolved from animals, we were not significantly different from animals. That was, of course, contrary to the basic Christian belief that humans are special and have an immortal soul. Theologians eventually responded with the theological idea that the human body evolves naturally but God infuses an immortal soul into each human being. Of course, the soul is not physical and cannot be scientifically detected. This did not convince the non-believers, but it did give the believers a way to understand how a spiritual, immortal soul in humans could be consistent with evolution. The concept of infusion provides an example of how the doctrine, i.e., the way the belief is worded, can be consistent with science. Note that the basic belief is that we have an immortal soul, but whether God infuses it or brings it about some other way, does not need to be part of the way that belief is worded in the doctrine.

Another example is the 7 days of creation compared to the 13.8 billion years of evolution. In this case, the theologians were able to argue that some parts of the Bible should not be taken literally because stories were used to give us God's message of salvation. The stories were not intended to give scientific details of God's creative process, which they obviously couldn't do at that time. Again, this did not convince the non-believers, but did give believers a way out of the issue. Overall, we believe that the Christian theologians did a very good job. Perhaps this is one of the reasons why Christianity continues to be one of the world's dominant religions.

Trying to understand evolution and resolve issues like those mentioned above has led to several different explanations about how God is involved in evolution. There is no consensus. All of these explanations are "on the table" and people discuss them. There are many current books on each topic. The most popular explanations can be categorized as follows:

Materialism: God is not involved in evolution because God doesn't exist. The only reality is physical reality. There is nothing spiritual or non-physical, no God, no angels, no human souls, etc. The universe is not designed nor is evolution guided.

Creationism: Some believers continue to doubt any scientific evidence that points to contradictions in what we know from a literal interpretation of the Bible. They stress the importance of being open to accepting God's word. We understood the position that it's not *how* you believe but *what* you believe that gets you to heaven. Nevertheless, Joe has always had real trouble trying to have a faith that contradicts anything we know from science. Ron was always a solid believer, knowing that there must be an answer, but he was curious and stayed with Joe on the journey.

God Driven, Natural Evolution: With this explanation, what we see is God's creative process. He is the creator and drives the evolutionary process from within, in ways that we see as being consistent with the laws of physics. God designed and fine-tuned the universe in the beginning and further intervenes during the evolutionary process, e.g., by infusing knowledge that allows the cells to know what to do. In other words, whatever we observe in science is God's creative process. There are no interventions that we see as violating the laws of physics.

Pierre Teilhard de Chardin, a Jesuit priest (1881-1955), was the first to present and make popular this explanation. He firmly believed that Christianity had to be made consistent with science and evolution or it would eventually become irrelevant. He thought it was necessary to make some "minor" changes to Christianity. By saying that evolution, as we observe it, is God's creative process, he guaranteed that Christianity would be consistent with evolution.

For Teilhard, and the many that have followed with different versions of his basic idea, this forces a key restriction on Christianity. That restriction is that there is no possibility for Adam and Eve to have been created in a Garden of Paradise in a state of "original justice" as is needed for the doctrine of original sin. In this state, Adam and Eve would have had a direct relationship with God, would not have been subject to physical suffering and death, and would have had complete control of their bodily desires. Instead, with natural evolution, they would have been created just like we are today: no direct knowledge of God, subject to suffering and death, and not in complete control of their bodily desires. In other words, with God Driven, Natural Evolution, there is no original sin event by the first humans, and we are not in a fallen state. With this explanation of creation and evolution, "original sin" is considered to be a general condition of physical evolution, not an event that actually happened in history. So, when God decided to create a physical universe, it had to have these kinds of "growing pains." An increasing number of Christian theologians today are favoring this explanation. However, most Christians, including us, believe that this answer to the question of how God is involved in evolution, gives up too much of basic Christianity in order to be consistent with science and evolution. We believe a better solution is needed.

God Driven, Natural Evolution with Some Physical Intervention: With this explanation, God is the Creator, and He guides the evolutionary process in a way that is generally consistent with what we observe. But there are a few exceptions and humans are one of them. In this view God wanted to have a direct relationship with humans and did not want us to suffer and physically die. He intervened and put the first humans, Adam and Eve, in the Garden of Paradise as discussed above. Unfortunately, they sinned and returned humanity to the natural state they would have been in had God not intervened. All following humans were and will be born in this "fallen state."

In the next chapter, we'll discuss how studying these explanations lead us to four very important, but unanswered questions. There is also a fifth question that we didn't think was related to God and evolution, but we were wrong.

Chapter 3

No Answers (1972)

There was the time when our families, without Joe and Ron, went to a drive-through zoo in Dallas. It was a carload. Sandy with her three children at the time: Bryan, Scotty and Stephanie, and Linda with her two boys: Gary and Danny. All five were under six years old. One of the residents they had a close encounter with was an ostrich. Everyone refers to ostriches as sticking their heads in the sand when they are scared or to avoid dealing with problems. But since they build their nests in the sand this common notion about them probably isn't true. In fact, the opposite was happening at the zoo. A large ostrich was pecking so hard on their car window that they were scared that he might break through the glass.

New paradigms of thought often need someone to break through the glass. Many folks like to give output and not entertain input. Most don't like challenges to their belief system which they have worked so hard to create. They can be like the proverbial ostriches that have their heads in the sand, scared of what might happen to their theories. There are many theologians who won't entertain new ideas on doctrinal problems and many physicists who insist that physical reality is the only reality in spite of experimental and theoretical evidence to the contrary. We've personally encountered both types over the years. We are trying to find some people in each discipline who can be more like the ostrich encountered at the zoo and be willing to listen to new ideas that can break through barriers.

In the following chapters we discuss the theological issue of how to reconcile an original sin event with evolution. We also discuss the scientific experimental and theo- retical evidence that the fundamental par- ticles of nature do not always behave in purely physical ways. It took us many years to realize that this theological issue and this scientific issue are related in a way that can only be understood within *a new paradigm* or way of understanding reality.

After thinking about our faith, what we knew from modern science and evolution, and the currently popular explanations about God's involvement in evolution, we realized that our journey had led us to some serious questions. We eventually identified four basic questions that needed reasonable answers in order to complete the journey. We could not find answers anywhere to any of these questions. Joe saw these questions as barriers to his faith and he felt that in the future there would be more and more people that would also see them as barriers. He argued that it was going to take some new ideas to break through these barriers. Ron's faith was not shaken, but he stayed with Joe on the journey. In this chapter we'll discuss these four basic questions. Actually, we had five questions, but we didn't think the fifth question was related to the issue of God's involvement in evolution. However, we later found out that it held the key to explaining our answers to all five questions.

Our five basic questions are:

1. Is there anything beyond the pure material universe?

Note that in materialism, fundamental particles like photons and electrons are purely physical and spiritual entities do not exist. We did not believe this made sense for several reasons. The fine-tuning argument, the apparent guiding of evolution, and a strong belief in God made materialism seem wrong to us. However, we could not develop a new and strong enough argument that we could use to take our position to others, especially those who did not agree with the fine-tuning argument and other evidence of design.

2. How are spiritual and physical interactions possible?

The materialists deny the existence of
spiritual, i.e., non-physical entities. Traditional Christians, and all others who believe
in God and life after death, certainly do
believe in the reality of spiritual entities.
However, they typically see reality as
consisting of two distinct types of entities:
physical and spiritual. The spiritual entities
are, of course, also non-physical. We agreed
with the reality of both types of entities but
could not understand any way that they could
interact. For example, if a soul is spiritual and
non-physical, how could it interact with a
body that is purely physical? It seemed to us
that if there was a direct connection, the soul
must have a physical element, or the body
must have a non-physical element. We did
not have any idea about how that might be
possible. We could not find anyone that had
an answer or even an insight for this question.

3. Why is there so much evil given a good
    God?

This question is hard to answer in all
religions. Many who have tried to believe in
God have given up because they could not
find an answer to this question that satisfied
them. We realized that before evolution, the
doctrine of an original sin of the first humans
was the best answer to this question for
Christians. But even then, it was hard to
understand why God would have wanted to
create a physical universe that would begin
with such a failure. Bishop Robert Barron has
recently pointed out that this problem has
long been called *mysterium iniquitatis* (the
mystery of evil). But more importantly,
evolution made the Garden of Paradise story
incompatible with science for many people.
This leads to a dilemma which seems to have
no solution.

With God Driven, Natural Evolution,
there is no original sin event. Therefore, God
is, one way or another, responsible for all the

evil in the world. That causes several problems for traditional Christianity because (1) it loses: the notion of original sin, that we are in a fallen state, and (2) that Christ is the Redeemer, saving us from that fallen state. It becomes a Christianity where evolution is God's creative process and Christ is more of a co-evolver, helping us get through the process. In other words, God either wanted, or could not avoid, creating us in a world full of evil, suffering and death. However, we did not feel that God creating us into this situation was consistent with the concept of a good and fair God. For most Christians, including us, this is exactly why we believe that some kind of an original sin event is the cause of the problems, rather than God or His process.

On the other hand, many believe that God Driven, Natural Evolution with Physical Intervention to create Adam and Eve in the Garden of Paradise is not compatible with evolution. This is such an important issue that it led to another question.

4. Can we reconcile original sin with evolution?

This became a key question for Christians when evolution was discovered, and it still is today. Original sin is the best explanation for the problem of evil. With an original sin event, God is good and everything He creates is good. A sin of pride, not physical weakness, becomes the cause of all the evil, suffering and death in the universe.

On the other hand, since an original sin committed by the first humans is inconsistent with evolution for so many people, something has to give. At the time Ron thought that this is a mystery that most Christians do not worry about right now but must face at some point in the future. Joe had difficulty in believing in an aspect of Christianity that was inconsistent with science. Both Joe and Ron believed that if Christians can't reconcile original sin with evolution, then eventually Christianity, in order to survive, will have to drop the belief in original sin as an event that happened. Joe did not like this, and Ron thought it was totally unacceptable, so they continued to work on this issue. Notice that everything else in Christianity would remain: God is good and Creator of the universe, Christ is the Son of God that was sent to save our souls, etc.

This kind of situation is explained very clearly by Brother Guy Consolmagno, SJ, Director of the Vatican Observatory, and Fr. Paul Mueller, SJ, Administrative Vice Director and Superior of the Jesuit community at the Vatican Observatory, in their book: *Would You Baptize an Extraterrestrial?* – Publisher: Image, an Imprint of the Crown Publishing Group, a division of Random House LLC, a Penguin Random House Company, New York, Pages 52-54:

This is often expressed in shorthand by the saying that God is the author of *Two* Books, the Book of Scripture and the Book of Nature. The traditional Catholic position is that these two books *cannot* disagree with each other, once they have been properly understood. Both books are written by one and the same author, God. And God does not disagree with God: "Truth Cannot Contradict Truth."

This means that, if there seem to be differences between science and the Bible with the respect to the creation of the universe and the creation of life on Earth, that merely means that we haven't yet managed to "read" or "interpret" the Two Books correctly.

So when science and the Bible seem to be in conflict, one possibility is that we don't have the science completely right; we haven't yet learned how to read the Book of Nature correctly. ...

The other possibility, when science seems to be at odds with the Bible, is that we haven't yet interpreted the Bible correctly; we haven't yet learned how to read the Book of Scripture correctly. Maybe we're reading the Bible literally, when we should be reading it figuratively—or vice versa.

The third possibility is that we're reading *both* books incorrectly—we have both the science and the scriptural interpretations wrong!

The disconnect between the literal interpretation of Adam and Eve as human beings and the scientific discovery of evolution has been around for a very long time with no apparent answer in sight. In our opinion, resolving this is the most important issue for the future of Christianity, and is needed for Christianity to maintain original sin as an event that happened, putting us in a fallen state.

We believe that it's been so hard to resolve because it is one of the third and most difficult of the three possibilities mentioned above. This means that changes are needed both in the way we interpret reality as discovered by modern physics and the way we interpret the story of Adam and Eve from

Scripture. We did not see at the time that both of these are necessary and that's why we didn't see the importance of the next question to our faith-science journey.

5. Can there be a logical interpretation of modern physics?

There has never been a logical interpretation of the strange behavior of fundamental particles. We did not think that this issue had any bearing on issue of God's involvement in evolution, but it was driving us crazy, and we couldn't do anything about it. We decided to leave it alone and turned to the other issues which we knew did affect the question of God's involvement in evolution. We had no clue that little "Danny" who was born in 1972 would become the Dan that made our team grow to three people. Not only that, but Dan went on to get a PhD from Caltech in physics and would later develop a reasonable answer to this question. Furthermore, it turned out that we actually did need his answer to complete our journey, but we didn't know that at the time. But we're getting ahead of the story, so we'll continue with this issue later.

In summary, our journey to understand how God could be involved in evolution in a way that is consistent with faith and science became a matter of getting reasonable answers to these five basic questions. However, it wasn't long before we realized that answering them to our satisfaction wasn't good enough to complete the journey. We also needed to have a deep enough understanding of the theological and scientific issues that we could explain our insights and answers in a way that made sense to others.

At this point in time (1972), we had an answer to only one of these five basic questions. We had concluded that materialism was wrong as discussed above. However, we had nothing new regarding our ideas against materialism that we could explain to others.

It took almost fifty years of searching and thinking before we realized why there were no really good answers anywhere for these five questions. The reason had to do with the two popular paradigms that people use to view reality.

The first paradigm is materialism, where the only reality is physical reality. In other words, physical equals real, or, if it's not physical, it's not real. Obviously, this paradigm can't shed any light on the questions that have to do with God and spiritual matters. Nevertheless, many feel that the successes of science have established science as a replacement for the belief in God and the spiritual realm. However, with materialism, there is no way to interpret the findings of modern physics in a logical way, much less the nature of consciousness or any of the answers related to non-physical or spiritual issues. These shortfalls in materialism, which were addressed in previous chapters, are covered again in the following chapters and, in more detail in our other book *The Fallen Angel Model.*

With the other popular paradigm, the one that allows for the existence of physical entities and spiritual entities, there is no way to explain how physical and non-physical entities could interact. Also, for many who believe in this paradigm, there is no way to reconcile the original sin of the first human beings with evolution as understood by science.

The point is that there is no possible way to get comprehensive answers to any of these five questions inside these two paradigms. It turns out that reality is more complicated than either of these paradigms allow for, and a richer view of reality is needed. It took us fifty years to understand this, but we can show you in this short book exactly how we answered all these questions to our satisfaction and how our explanation of creation and evolution can make sense to you. However, in 1972 we did not yet have the insights we needed to build upon. We were reading and discussing both science and theology. We had the questions, but we didn't have the answers.

Chapter 4

Danny on the Rocking Horse (1975)

Throughout the early 70's Joe and Ron were working in different directions.

Joe: I became interested in philosophy in college. I noticed that the great philosophers before modern science, especially like St. Thomas Aquinas, had asked and gotten answers to the big questions about the nature of reality and how we should live our lives. My problem with them was that their answers weren't based on what we now know from modern science. So, I started looking at the modern philosophers. I found that they had long since given up trying to answer these questions. Then in a flash, I saw that the history of philosophy itself could be seen as an evolutionary effort to get answers to these big questions. I'll never forget that moment. I remember the room I was in, and I remember little Danny, who was 3 years old, rocking back and forth on a rocking horse in the living room. Eventually I completed the effort and published *The Philosophy of Conscious Energy—Answers to the Ultimate Questions* in 1993 (and later republished it as *Conscious Energy and the Evolution of Philosophy* in 2021). That book provides the history of philosophy as an evolution of thought and introduces the Philosophy of Conscious Energy which provides purely philosophical answers to the two big questions based on what we know from modern science.

I thought that since I was working on pure philosophy, it had nothing to do with our journey trying to understand God's involvement in evolution. It turned out that some of the key ideas that I learned were not only needed to develop the Philosophy of Conscious Energy but were also needed to take the next step in our journey.

The key ideas are discussed below.

Our understanding of Teilhard de Chardin was that he saw our universe as one of energy in various states undergoing a transition of increasing complexity and consciousness guided by God. As energy changes state it can have very different properties. A simple example of what we mean by "change of state" is what happens when water freezes. It is still the same underlying substance but has very different properties when it becomes what we call ice. When it melts, it returns to its previous state. Another example is the immortal soul in humans, although ultimately created by God, in Teilhard's view can be thought of as a state of energy and does not have to be directly inserted by God. It can evolve and change state from matter and energy at lower states in process guided by God. Of course, we immediately saw that this was consistent with Einstein's famous discovery that radiation and mass, or matter, are two different states of energy with very different properties. The ideas that energy is fundamental in our universe, can change states, and have very different properties in different states are very powerful concepts. Thinkers before modern science did not have these concepts and, in our opinion, these concepts have not been used enough by modern thinkers.

At this point we began to consider the idea that the human soul and angels are perhaps some kind of "spiritual energy," i.e., non-physical but self-conscious and having the ability to be in a close relationship with God. We also thought of the early universe as being non-physical, i.e., containing no matter, but not in a close relationship with God.

Then we began to see how a non-physical soul might somehow be able to interact with a physical body since both are energy in different states. However, even with both as energy, they are in different states, and we couldn't see an interface between something purely physical and something purely non-physical. Nevertheless, considering that both are energy we felt we had a reasonable answer, or at least an insight, to this question.

In summary, energy as an underlying substance in our universe and the concept of change of state are two key ideas that we needed to allow us to understand how physical and spiritual interactions might be possible. However, the answer was not detailed enough that we could explain it to others, although we tried many times.

In the next chapter we'll discuss a big new idea that allowed us to see a way how God could be involved in evolution. It is a way that had never been on the table before.

Chapter 5

The Ice Storm in Dallas (1979)

Joe: Back when we were much younger, Ron and I loved to hunt, especially for ducks and geese out in the swamp and bring them home for dinner. On one hunt, we drove a couple of hours in the dark, early morning. Ron and I walked a long distance through a stinky marsh, got settled, and set up the decoys. We were all full of expectations for a great hunt. Almost as soon as we got ready, Ron said "Let's go home; I'm really sick." "What?" I yelled. It was a gut-wrenching disappointment to have to leave this perfect place where we had just arrived. It took forever to get out of that swamp because Ron, who weighs about 50 pounds more than I do, could hardly move, and I couldn't carry him. Ron had the dry heaves and kept falling into the stinky sulfur fumed marsh. This happened many years before cell phones, and we had no choice but to keep moving. We eventually made it back to the car.

Moral: If you want something bad enough, you need to keep going through disappointments. Meanwhile, we kept moving through disappointments on our science and faith journey.

In early 1979 the pieces for one of the original ideas that we needed had been in place for a few years, but we hadn't put them together yet. We had been through many disappointments, but we hadn't given up. The pieces were (1) energy can be thought of as the basic stuff of the universe, (2) it can change state, going from the physical to the non-physical and vice versa, (3) the universe is evolving under God's guidance to increasing levels of complexity and consciousness, (4) St. Thomas Aquinas' explanation of creation had been good for centuries and was completely consistent with Christianity.

However, he did not know about evolution and change of state. (5) Teilhard's view of creation was completely consistent with science and evolution, but he did not speak of angels, and he had ruled out original sin as an event that happened in history.

The new idea discussed in this chapter later came to be known as The Fallen Angel Model (FAM), but it didn't start out with that name. This chapter covers the story of exactly how these pieces suddenly came together for us. In this chapter, you will discover the idea just exactly the way we discovered it.

The concept of fallen angels as the basis for the beginning of the universe occurred to us in the winter of 1979 during a rare ice storm in Dallas, while we were discussing creation and evolution. The basic question we were addressing was: Can we reconcile original sin with evolution? It was a wonderful way to pass time in an ice storm.

We had been working on this question for many years and been down many blind alleys. The ice storm gave us the opportunity

to try another alley. We figured that St.
Thomas Aquinas and Teilhard, when taken
together, had covered all the issues in science
and theology, but neither alone had the
complete story. So, Joe suggested that we put
their ideas on a piece of paper and see if that
might lead us to an answer to this question. A
reproduction of that drawing is shown below
in Figure 1: Comparing St. Thomas Aquinas
and Teilhard de Chardin.

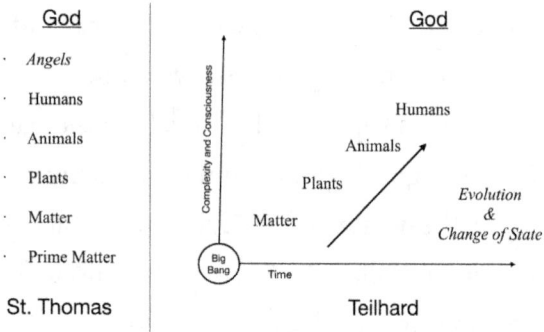

Figure 1. Comparing St. Thomas Aquinas and
Teilhard de Chardin

If you look at this figure, you'll be able to
see exactly what we were thinking. We put St.
Thomas' vision of creation on the left of the

vertical line and Teilhard's on the right. Of course, St. Thomas didn't know about evolution, but he did see creation as a hierarchy of being and he considered the angels to be part of our universe:

> …the angels are part of the universe: they do not constitute a universe of themselves; but both they and the corporeal natures unite in constituting one universe. — *Summa Theologica,* Volume I, Question 61, Article 3

On St. Thomas' side, God, the Creator, is shown at the top, angels are shown just under God in the hierarchy, then humans, animals, plants, matter, and something he called "prime matter" is shown at the bottom. Of course, St. Thomas did not have the concept of evolution, i.e., creation over time, so each different kind of creation, starting with angels and ending with prime matter, is shown as a bullet item. Note that the hierarchy is one of importance or closeness to God, not necessarily the order of their creation in time.

However, St. Thomas did have the concept that God creates in two ways. St. Augustine had pointed this out centuries earlier based on his very insightful reading of Genesis. (Word on Fire Video — Wonder: Seeds of Life.) St. Augustine had noted that: (1) Sometimes God creates new things directly out of nothing, e.g., "Then God said: Let there be light, and there was light." — Genesis 1:3, United States Conference of Catholic Bishops (USCCB) Bible. (2) Sometimes God creates new things from things He had previously created, e.g., "Then God said: Let the earth bring forth vegetation: every kind of plant that bears seed and every kind of fruit tree on earth that bears fruit with its seed in it. And so it happened: ..." — Genesis 1:11, USCCB Bible. The point we're making here is that although St. Thomas did not know about evolution, he did know that God creates new things from things that He had already created, which, of course, is consistent with evolution.

We put Teilhard's vision of creation and evolution on the right side of Figure 1. He also believed that God is the Creator and again, God is shown at the top. However, Teilhard believed that evolution is God's creative process and everything God created was accomplished through evolution. Therefore, time is needed and is shown on the horizontal axis. Over time, more and more complex and finally, self-conscious creatures evolved. The vertical axis is a measure of this, i.e., increasing complexity and consciousness. The entries are almost the same for St. Thomas and Teilhard except Teilhard didn't have angels, and of course Teilhard's vision was an evolution over time. After the Big Bang, radiation and elementary particles began to form. The pre-matter in the early universe reminded us of the prime matter described by St. Thomas. Eventually everything in our universe evolved: matter, plants, animals, and human beings.

In order to better distinguish between St. Thomas and Teilhard, we put the word *Angels* in italics on St. Thomas' side because Teilhard didn't discuss angels and we added the words *Evolution & Change of State* on Teilhard's side in italics because St. Thomas didn't know about these concepts. One of Teilhard's most important ideas was that matter, evolving under the influence of God, could change state and become spiritual. Given this idea, we believed it was reasonable to assume that under certain circumstances some spiritual entities could devolve or "break" into non-physical entities and then transform into physical entities.

Of course, from Scripture, tradition, and doctrine we know that God created the angels and gave them free will. We also know that some of the angels sinned and were thrown out of heaven.

Given all of this we believed that all the pieces we needed were on this page but were not organized or linked properly. We then began to think about how to link their theories to get an idea about what might have happened. Eventually we saw the "fall" as downward movement and we drew an arrow straight down from the word angels.

Then in a flash we saw the "missing link" between the two models and drew another arrow down from the word angels on the left toward the right—right to the Big Bang. Both arrows are shown in Figure 2.

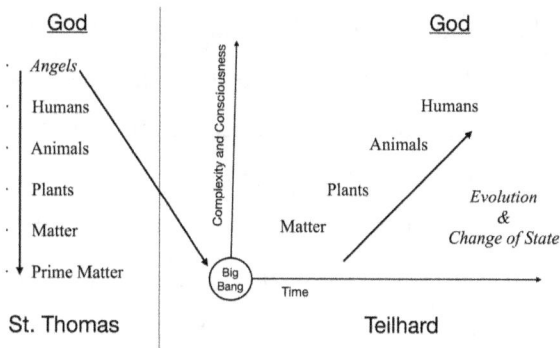

Figure 2. Showing the Fall of the Angels being linked to the Big Bang

Soon we had the concept that the matter and energy of the beginning of the spacetime universe could be thought of as "unordered" or "broken" spirit. It was a moment we will never forget, and honestly, we have never looked at the universe the same since then.

This model of creation is consistent with science which tells us that the very early universe was non-physical. There was no matter and even the laws of physics weren't fully formed. This fits with the concept that some kind of spiritual energy fragmented, i.e., became "broken" and started to devolve/evolve. We called it the Angel Fragmentation Scenario (AFS). Note that we are using "spiritual energy" in the way many theologians use the word "supernatural" to represent entities in the realm with God.

This idea also provides a new insight for the fine-tuning of the constants of nature. Instead of the constants being individually designed, they are like broken pieces of a glass vase which appear to be fine-tuned to fit together nicely. However, they are just the result of what happened when something spiritual, which has a very high degree of order, changed to something non-physical and started to devolve and evolve.

With this explanation about how God is involved in evolution, everything that God creates is good. All evil, suffering, and death are the result of the fall of spiritual creatures who were with God in heaven. In other words, they were in a state of original justice which means close to God, not subject to physical tendencies, or any suffering or death. It is also consistent with the belief that we are born into a fallen state, but in this case, the whole physical universe is born into a fallen state because the fall occurred before the observable universe came into existence. With this explanation, Christ can indeed be viewed as the Redeemer of humans who evolved from a fallen state.

A picture of the actual drawing we used is shown in Figure 3.

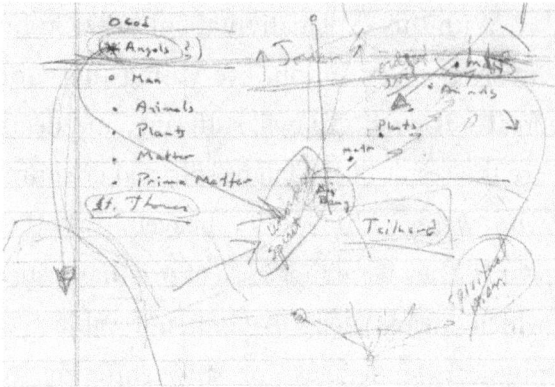

Figure 3. A copy of Joe and Ron's original
drawing comparing St. Thomas and Teilhard
which led to the Angel Fragmentation Scenario

Figure 4 shows a copy of the photograph
that Linda took of Joe and Ron right after we
got the idea about the fall of the angels being
linked to the beginning of the universe. It
really was a moment that changed our lives,
and we'll never forget it.

*At this point we had a new explanation for
how God is involved in evolution.*

Figure 4. Joe and Ron,
just after we got the fallen angel idea in 1979.

With this explanation, God did not
directly create the evolving energy of the
beginning of our universe in that state. The
sin of the angels caused a change of state to
some of the beautiful and stable spiritual
energy that God had created. God chose to
work with that evolving energy in ways that
we see as consistent with the laws of physics.
This is a short explanation of the insight that
we had then, but that insight was not
complete enough to be convincing to others.
There were too many gaps or missing details.
We will provide a better explanation later as

we deal with the rest of the five basic questions.

So, although we tried very hard, we couldn't successfully explain these ideas to many people. The traditional Christians had trouble accepting the idea that the fallen angels "fragmented" and lost their consciousness because current doctrine says that all angels, even the fallen angels, are immortal. Many also wanted to keep the doctrine of original sin as exactly as currently worded with "Adam and Eve" as humans, i.e., they did not want to reinterpret or extend the doctrine to angels. Another stumbling block was that most people had the understanding that physical and spiritual entities are separate. They did not see everything in the universe as different forms of energy that can change state from spiritual to physical and vice versa. So, even those who believed in the existence of spiritual entities, could not understand how something spiritual could become something physical.

Nevertheless, we realized that this fallen angel idea was much less disruptive to Christianity than the alternative of no original sin event, no fallen state, and Christ is co-evolver, not a Redeemer. Of course, materialists would not accept even the existence of angels because angels are non-physical and, therefore, not part of the materialistic paradigm of reality.

In a separate issue, we still didn't have any way to logically understand how the fundamental particles of nature behaved, but we were still thinking that this had nothing to do with the God and evolution related questions, and we were wrong.

As of 1979, we were able to take a couple of new steps based on the Angel Fragmentation Scenario (AFS), which is the early version of the Fallen Angel Model (FAM). These steps allowed us to have answers to two more of our questions: the problem of evil and reconciling original sin and evolution. Nevertheless, we did not have a deep enough understanding of them to be able to express our explanations in a way that made sense to others. That requires a new paradigm or way of looking at reality.

Here's another interesting twist to the story of this journey. We didn't know it at the time, but there was a much easier way to come up with the idea that the fallen angels had something to do with the beginning of the universe. We could have used the concepts of energy and change of state with a little common sense.

The common-sense argument goes like this. Start with the question: How can we reconcile original sin with evolution?

If there's no way to fit an original sin event inside of our spacetime evolving universe, then there are only two possible answers: (1) It didn't happen at all, which is what an increasing number of theologians are saying or (2) there could be a sin outside of spacetime that is related to our spacetime universe. The fall of the angels is a well-known event in both Scripture and theology. We didn't see this at the time, but it certainly looks obvious now. Of course, it's not obvious unless you have the idea that energy is the basis of fundamental reality in our universe and the idea of change of state.

Chapter 6

Dan in Graduate School (1996)

Ron: Dan Pro has a PhD in physics from Caltech. He chews his food very slowly, really thinks things through, and he concentrates even harder than Joe. Dan has always needed his faith to be consistent with his science. It made us so happy when he agreed to be an author of our previous book (FAM) because we needed his clear thinking and physics knowledge. Dan once got involved in a "dead end" project while he was working on his doctoral thesis. He had come up with an experiment that would prove light, under some circumstances, could go faster than c (299,792,458 m/sec in a vacuum). Dan went to almost all the physics professors at Caltech and none of them could find any errors in what he was proposing. So, he worked hard on it, both in the lab and on paper for about six months. Finally, someone saw the flaw in the paper. He went back to his professor and told him what happened, and the professor was glad the issue was resolved. His professor realized that Dan is a real "truth-seeker." Dan had to know if his proposal was right or wrong and why. He didn't care what the answer was or who solved the problem. That determination ultimately led to the key breakthrough we needed to complete our 50-year journey and that story is covered in this chapter.

In 1996 Joe was thinking of ways to explain our fallen angel idea but he wasn't having much luck. In 1993 he had introduced the fallen angel idea in the latter pages of his philosophy book *The Philosophy of Conscious Energy*. Ron was still studying physics and thinking of how we might get the idea of the fallen angels more in line with traditional Christian doctrine.

Meanwhile Dan, at 24 years old, had become a member of the Joe and Ron journey to try to understand how God is involved in evolution. He was also very busy in graduate school studying applied physics at Caltech. Of course, Dan had also run into the wave-particle duality and the mysteries of fundamental particles. Unlike most physics majors, he really didn't like being told he had to choose between illogical interpretations and "you can't ask the question" responses. He finally asked his professor if he could take some time off his graduate work to see if he could make some sense out of the mystery.

Unlike Joe and Ron, Dan had the benefit of several classes in quantum mechanics (QM), which is a modern physics theory that predicts the outcomes of fundamental particle movement. A very non-technical explanation of the problem is that the fundamental particles in our universe sometimes behave as particles, sometimes as waves, and move in ways that can't be physically described. This has been proven many times in experiments. Since physics is all about predicting what we can physically measure, the accepted position when doing physics is that "only the physical is real."

Dan: I did not challenge the physics, for good reason; QM works just fine. It predicts the statistical outcomes of experiments. However, it does not predict exactly where each particle will go or the trajectory it will take. The simplest example is a fundamental particle going through two slits that are close together. The equations of QM predict statistically where the particle will hit the detector. With repeated experiments, it has been proven over and over that the equations work. The problem is: how can a particle go through both slits, but land in only one spot at the detector behind them? QM doesn't have any explanation for how the electron cloud of energy "collapses," which means being spread out in space one instant, and then completely disappears and collapses to a localized spot where a particle interaction was detected.

It was during this period that my dad and I began to meet for lunch with a group of PhD's in the Caltech/JPL community who were interested in the various interpretations of QM. We called ourselves the Quantum Lunch Bunch. Many lively discussions gave us a chance to interact with some very smart people who held positions that just didn't make sense to us. The process made us more determined that there must be a reasonable interpretation and we needed to find it.

For over a hundred years, the consensus among physicists has been that there is no logical way to understand what happens in between the slits and the detector.

So even though there is a non-physical behavior going on, I was insis-tent that reality still must be logical. All the other interpretations of quantum mechanics either abandon logic altogether or abandon reality itself. That's because they all insist that quantum mechanics is the complete understanding of physics. It did not seem reasonable to me that we can just assume that we live at the time when the nature of fundamental particles is completely understood. So, I concluded that scientists haven't learned enough theory yet. In my opinion QM theory is not wrong but falls short because it doesn't explain HOW a wave of energy collapses. Some other undiscovered theory will come along and explain it, most likely leaving intact the parts of QM theory that are in agreement with experimental results.

What I ended up doing was write up a unique interpretation of quantum mechanics that keeps the reality of the wave function but allows the behavior of fundamental particles to be non-physical. I called this the "ProWave" interpretation of quantum mechanics. It's an obvious double meaning of "Pro." It favors the wave over the particle as the explanation of reality and it's Dan Pro's idea.

The "particle" may not be purely physical, but it's real. In other words, Pro-Wave provides a logical answer, but not a purely physical answer. To us, the word "physical" means that particles follow trajectories described by physical laws of motion. (We discuss this in depth in Chapter 8.) As for waves, quantum mechanics describes physical outcomes and interactions, but not paths taken. I tried to publish my paper, but it was rejected because "It's not original physics." Of course, I knew I was not suggesting new physics; I was just suggesting a new interpretation of what physics had discovered. We have always wondered if that's the real reason why it was rejected. Perhaps the real reason is that ProWave claims that the "physical" world is not completely physical at the level of the fundamental particles, like photons and electrons. That thought is vehemently

disliked by the culture of popular physicists, especially those who are materialists. For the materialists, physical
means real and real means physical. In
other words, the suggestion that a particle
in nature could exhibit a non-physical
behavior was just not in their paradigm,
or view of reality. It turned out that we
needed a new paradigm, or way of viewing reality, before we could explain Pro-
Wave to others.

A writeup of the original ProWave
paper is included as an appendix in our
book *The Fallen Angel Model*. In that
paper, I provide a verbal explanation of
the double-slit experiment and many
other experiments. I also provide the
equations for all the examples. We also
have another appendix in that book
where you will also find a deep insight
into why imaginary numbers are necessary to explain the non-physical compo-
nents of the behavior of fundamental
particles. Actually, we believe that it

would be more insightful to call them "non-physical" numbers rather than "imaginary" numbers. Of course, when measured, the particles are in a purely physical state and that is consistent with solving the equations for the states where the non-physical aspects are zero.

Of course, this issue is more complicated than expressed above, but the short answer is that ProWave, written in 1996, offers a way to have a logical, even if not a purely physical, understanding of how the fundamental particles really behave. Joe and Ron were thrilled to have this.

ProWave also marks a key milestone in the journey. For the first time we had answers that we were satisfied with for all of our five basic questions. Although we tried repeatedly, we still could not explain any of our answers in a way that others could see as valid.

We still thought that ProWave had nothing to do with the question: How is God involved in evolution? We didn't yet see that ProWave would pave the way forward by allowing us to explain our fallen angel idea to others.

Chapter 7

The Double Tire Blowout (2006)

Ron: Joe Pro is also known as Tex Fiddler in the world of Cowboy Action Shooting where he is a three-time Senior World Champion. He is always wearing a black cowboy hat, black boots, a large belt buckle, and there is a pocket watch in his leather vest. On special occasions, he wears the watch that belonged to his great-grandfather and that watch has a small gold crucifix hanging on the chain. Joe's Catholic family goes a long way back in time. He wears his hat when he leaves the house, even if he is going to the mailbox. Some might say he is a bit eccentric. But he is just Joe Pro, the author of 5 science-faith books. He can really focus on a problem, but he can only do one thing at a time, and he's hard-headed. Contrast this with me. I am all over the place, but also hard-headed. It makes for a good team. Oh, I almost forgot, Joe Pro shaves with a straight razor.

A "feud" developed between Joe and Ron, and it went on for decades. Joe felt that fragmented angels made perfect sense. He figured that since God is the ultimate in self-consciousness, it made sense that being apart from God would mean the opposite, i.e., no consciousness. Ron argued that, although logical, fragmentation and loss of conscious-ness was not necessary for our ideas to be complete and gain attention. More impor-tantly, Ron argued that there is too much scripture and theology that addresses "fallen angels" or devils and we can't just ignore that. Then suddenly the issue boiled over into an incident that we call the Double Tire Blow-out.

Joe: Ron and Sandy had come to California for a visit, and Ron and I kept discussing fragmented vs. fallen angels. I was driving Ron and Sandy to the airport in the rain and Ron was driving me crazy. The argument got heated, I got distracted, and missed a turn. Then in frustration I tried to make a fast turn but ended up sliding sideways into a curb. Both tires on the driver's side of my pickup truck hit the curb at the same time, blowing out both tires. We had to call a cab to get Ron and Sandy to the airport and I called Linda to come and pick me up. The truck had to be hauled in for some new tires. The double tire blowout ended the discussion for that day. I put on my full-length black duster, cowboy hat, waited in the rain, and cooled off.

Meanwhile, I was still thinking about how we could better explain our fallen angel ideas to other people. In 2012 I published a free e-book intended primarily for college students titled *How to Believe in God and Science — In Three Easy Steps.* The book provided a comparison of all the possible ways that God could have created the universe with all the possible obstacles to believing in God. The idea was that this comparison would show that the fallen angel idea provided the only explanation that could overcome all the obstacles to faith. However, it turned out that such a comparison, although logical, was complicated to read, resulting in the "easy steps" becoming "hard steps." Furthermore, I did not yet realize that successfully explaining the fallen angel idea required introducing a new paradigm or view of reality. Even people who believe in God and angels just can't fit this concept in a world view where the spiritual and the physical are two separate realities.

Ron: In 2006 I was ordained a deacon in the Catholic Church. The concept of the angels fragmenting, which had always bothered me, became more of an issue. But the alternative, i.e., that there was no original sin event, bothered me even more. One day I thought that maybe the fallen angels didn't fragment and lose their consciousness but lost some kind of energy that did fragment. If that were true, a fallen angel model would be consistent with the Scripture, tradition, and doctrine which tell us that even fallen angels are immortal.

It has always been believed throughout history that the fallen angels were "darkened" and lost something when they fell. Joe agreed that it made sense to believe that the angels lost some energy, perhaps some conscious energy, some power, and certainly their love of God. Since Scripture is full of references to fallen angels tempting us and many theologians refer to all angels as being immortal, Joe finally decided he should not insist that they fragmented. It took some humility and a little bit of common sense to reach that conclusion.

Given this decision, we changed the title and substance of the Angel Fragmentation

Scenario (AFS) to the Fallen Angel Model (FAM). We now had a solution to the problem of evil that was consistent, not only with evolution, but also with the Scripture, tradition, and the doctrine that all angels are immortal. Most importantly, we still had the key concept that the fall of the angels released energy that became the beginning of our spacetime universe.

The Fallen Angel Model (FAM) also gives insights into two mysteries related to the initial moments of creation as described in Genesis. The Bible begins: "In the beginning, when God created the heavens and the earth..." — Genesis 1:1, the United States Conference of Catholic Bishops (USCCB) Bible. In the footnote of this version, the bishops point out that: "Until modern times the first line was always translated, 'In the beginning God created the heavens and the earth.' Several comparable ancient cosmogonies, discovered in recent times, have a 'when...then' construction, confirming the translation 'when...then' here as well. 'When' introduces the pre-creation state and 'then' introduces the creative act affecting that state. The traditional translation, 'In the beginning,' does not reflect the Hebrew syntax of the clause."

We understand the bishops' comment to mean that adding the word "when" gives a more correct translation for the first sentence of Genesis and this implies that there is a pre-creation state. Without the word "when" it sounds like "In the beginning" refers to the first thing God ever did. With the word "when" it sounds like this is the story of what happened *when* God decided to create the heavens and the earth. We believe that God was active before He created the heavens and the earth and during that time, He created the angels and some of them fell.

Many people believe that we cannot talk about anything "before" the Big Bang, or however our observable universe began, because time began when spacetime began. But we really should say that "time as we know it" began with the beginning of our spacetime universe. The key phrase here is "time as we know it." Our faith tells us that God existed before He created the observable universe and that He created the angels outside of spacetime. This subtle point of inserting the word "when" is one of the very few places where anyone speaks, or even hints, of anything before the beginning of our universe. It's very interesting that we have found in modern times an ancient reference to the implication of a state that is before the beginning of our observable universe, and we have found it right at the beginning of the Bible.

The second mystery that FAM provides an insight into is in the mystery referred to in the remainder of that first sentence. That sentence continues by saying "and the earth was without form or shape, with darkness over the abyss and a mighty wind sweeping over the waters. Then ..." — Genesis 1:2, USCCB Bible. The bishops point out in a footnote, "God brings an orderly universe out of primordial chaos merely by uttering a word." This immediately begs the question: Why would God begin His creation of the physical universe in a state of "primordial chaos?" The Bible does not address this question, theology and tradition do not address this question, and, of course, science does not address this question. With FAM, we have the insight that God created the angels with free will in a perfect, spiritual state and their sin resulted in the chaos of the early observable universe. God then begins His creation of the "heavens and the earth" by choosing to bring forth as much goodness as possible, given the consequences of the fall of some of the angels. Of course, this also provides a new answer for the mystery of evil, i.e., why God would create a physical universe born in chaos and evolving full of evil, suffering and death.

The one theological issue remaining with FAM was that it is based on a sin of the angels whereas the current wording of the doctrine of original sin is based on a sin of the first humans. However, the basic concept is the same in both cases: God created spiritual creatures with free will, some sinned and it had a massively bad effect on our universe from that point on. If we are correct and it becomes widely accepted in the future that an original sin of the first humans is inconsistent with evolution, then Christianity will have to drop the notion of original sin as currently understood. But it won't have to drop the basic concept of original sin as an event that happened in history, that we are in a fallen state, and that we need to be redeemed by Christ. We believe, as is the case for seven days of creation, the literal interpretation that humans committed the original sin is not essential to the basic concept of an original sin event.

It's important to distinguish between a basic truth and how doctrine presents that truth. This was clearly pointed out by Pope John XXIII on the opening day of Vatican II:[1]

---

[1] Available at https://vatican2voice.org/91docs/opening_speech.htm

The substance of the ancient doctrine of the deposit of faith is one thing, and the way in which it is presented is another. And it is the latter that must be taken into great consideration with patience …

Unfortunately, this kind of confusion where something that is unnecessary is added to the wording of a doctrine has occurred in the past. The classic example is what happened with Galileo. The essential belief regarding creation and redemption had always been that God created human beings to be in His image and that Christ came to Earth as our Redeemer. The phrase "in His image" does not imply that God has a body, but rather that we have a spiritual soul that is self-conscious and can live beyond the death of our physical bodies. This is the basic Christian belief about creation, and it has never changed.

However, over time, some theological thinking got added into the way the doctrine was worded. In those days, the center of any object was considered to be a special place. Since humans were the most important and redeemed by Christ, it seemed obvious that God would put the earth in the center of His creation, and everything would revolve around it. This also agreed with the view of the universe as it had been developed and understood for centuries. Therefore, it was reasonable thinking for the time, but it was not needed in the doctrine, and, in retrospect, should not have been included.

When Galileo discovered and published that the earth revolved around the sun, he did not think that this should call the whole Christian belief of creation into question, but many did think that. The key point is that the earth being in the physical center of the universe is not essential to the belief that God created us, we are in His image, and He sent His Son to redeem us. It took some time, but eventually the wording got straightened out. The point is that the basic belief was always correct, but the way it was presented and dis-cussed in the doctrine had unnecessary additions that were later proved to be incorrect. Changes to the wording of doctrine like this are rare, but it is very important that it be done when it's needed. It is our opinion that it has happened again, this time with evolution and original sin.

Original sin is extremely important because it is the only completely satisfying solution to the problem of evil, i.e., a good Creator and so much evil, suffering, and death in the world. For example, we don't think it's logical that God would create us with a strong physical tendency to sin and then punish us for eternity in hell when we sin. We believe that Christians need to have a version of original sin that is completely consistent with science and evolution. In 1985, Cardinal Ratzinger (later Pope Benedict XVI) said: "...the inability to understand 'original sin' and to make it understandable is really one of the most difficult problems of present-day theology and pastoral ministry."[2]

We believe that, sooner or later, Christians will have to choose between the Fallen Angel Model and some version of creation without original sin as an event. We think that when this happens, Christianity will choose FAM because it retains the basic concept of original sin as an *actual event,* putting the whole universe in a fallen state.

At this point we became very comfortable with FAM because we believed it was completely consistent with Scripture, doctrine, and tradition regarding the immortality of all the angels and the basic concept of an original sin event. We also believed that it was completely consistent with science.

However, we were still having trouble explaining FAM to both believers and non-believers because of three reasons. (1) We

---

[2] *The Ratzinger Report,* p. 79, available at https://www.americamagazine.org/issue/350/article/evolution-evil-and-original-sin

couldn't really explain exactly what the fallen angels lost; (2) We hadn't really distinguished a clear difference between "spiritual" and "non-physical" and how something spiritual could become something physical; and (3) We couldn't adequately explain how something non-physical could interact with something physical. These are difficult issues and resolving the second and third issues required a completely new paradigm, or way of looking at reality. We didn't have that yet.

So, the change from fragmented angels to fallen angels, whose fall released some kind of energy, put us on much more solid theological grounds. However, it was still not enough to really explain the complete idea to others. Therefore, in 2006 our journey status remained the same as it was in 1996. We had answers to all five of our basic questions and they were getting better, but not to the depth that we could explain them in a convincing way to others.

In the next chapter we'll discuss the next step taken which was the development of a new paradigm or way to view the fundamental nature of the universe.

Chapter 8

Back to Texas (2018)

Joe retired from the Jet Propulsion Laboratory in early 2016 and it became clear that he needed to live near Ron in order to write a book explaining our ideas. By early 2017 Joe and Linda had moved back to Texas. It was at this point that Joe, Ron, and Dan began to get serious about trying to write these ideas in a book. The first draft title of the book was *The Blur* which was supposed to represent the strange, non-physical behavior of the fundamental particles discovered by modern physics.

Looking at the findings of modern physics, it's clear that the fundamental particles of nature behave in ways that everybody agrees are strange. But what is "strange?" The majority of people who study physics just stop there. Dan went beyond that with Pro-Wave and said that we should just accept the fact that "strange" really means "non-physical and real." Most physicists, whether they are materialists or not, don't want to hear that. Ron found more and more examples of particles and fields in modern physics experiments and theories that have attributes and behaviors that are non-physical. For example, all photons move at the speed of light the moment they are created. Purely physical things can't do that.

Soon we realized that what we were thinking was exactly what everybody didn't want us, or anybody else, to think. That attitude reminds us of a quote attributed to George S. Patton: "If everyone is thinking alike, then somebody isn't thinking." Our thinking was that the universe contains entities that are not purely physical. We feel this is clearly demonstrated by physics experiments and the strange "particles" in the equations of quantum mechanics. Ron pointed out that if we wanted to present this conclusion to others, we needed to be more precise with what we meant by "physical" and "non-physical."

Dan took the lead in the next step, and we developed more precise definitions. The following few paragraphs are taken from our first book, *The Fallen Angel Model*.

However, before proceeding, we need to be more precise about what we mean by the terms: physical, not-so-physical, and non-physical.

By ***physical*** we mean something that exists in reality that we can, at least in principle, detect, measure, and that only behaves in ways that can be described by the laws of physics.

The things we experience in everyday life provide simple examples of what we mean by physical entities, things like rocks, the air we breathe, and the moon.

By ***non-physical*** we mean something that exists in reality that we cannot detect, we cannot measure, and that behaves in ways that cannot be described by the laws of physics. ...

By ***not-so-physical*** we mean something that exists in reality that has at least one physical aspect and at least one non-physical aspect.

For example, we would call an entity not-so-physical if we can detect and measure it, but it has behaviors that cannot be described by the laws of physics. Fundamental particles such as the electron are examples of the not-so-physical. They are detectable and measurable but behave in ways that cannot be described by the laws of physics. For example, physics cannot describe the electron's behavior between the slits in the double-slit experiment and the point of impact on the detector. ...

Notice how the fundamental particle going through the double slit *changes state,* going from a wave of energy, and then collapsing, in a way that can't be physically described, to a physical particle that can be measured at a precise location.

It took some time, but we finally realized that this is a completely new paradigm, or way of looking at the fundamental nature of our universe. We called it the Not-So-Physical or NSP Paradigm.

As previously discussed, the two currently popular paradigms, or ways of explaining the fundamental nature of our universe, are (1) materialism, where physical reality is the only reality, and (2) the view that both spiritual and physical entities exist. In the second paradigm, the spiritual entities like angels and human souls are real but are not objects of study by science. It was clear to us that neither of these paradigms was rich enough to explain the nature of reality as discovered by modern physics. Neither paradigm had a concept of not-so-physical and neither had the concept that entities could change state from physical to non-physical and from non-physical to physical.

In summary, with the NSP Paradigm our universe consists of entities that are non-physical, not-so-physical, and physical. With the right conditions, some entities can change state from non-physical to not-so-physical and from not-so-physical to physical, and vice versa.

The NSP Paradigm, together with Pro-Wave, allowed us to explain how modern physics exposes materialism as inadequate and how ProWave provides a logical, but not physical, interpretation of modern physics.

In the next chapter, we explain how this new paradigm allowed us to resolve the remaining issues that were blocking our ability to explain our answers to the rest of our basic questions.

Chapter 9

The Final Steps of the Journey (2021)

Joe: In the old days, Ron would often go into the garage to light up a cigarette. I'd go with him, and we'd use that opportunity alone together to talk about philosophy in general and Teilhard de Chardin's idea of the "within" in particular. I suggested that he should read Teilhard's book *The Phenomenon of Man*. He did and he was struck by the idea of the law of increasing complexity and consciousness. He could then see how the periodic chart had underlying principles that guided the structure of atoms and molecules. Later Ron studied some of the physics and chemistry principles that were responsible for how the electrons and protons would go together to fill up the chart. But he didn't understand where these principles came from until FAM. Yes, the proton/electron forces of quantum mechanics can explain the nature of the chart, but going deeper, why is reality so organized like this? We see these principles as a result of the "broken spirit," baked in at the moment of the creation of spacetime. They came out of the Superforce which is the way physicists describe certain aspects of the very early universe. At that time, even the fundamental forces of nature had not become separated as they are now. Eventually, Ron felt that if FAM was correct, we should be able to see traces of the "broken spirit," non-physical reality in the physical universe today. But what is "broken spirit?"

Later on, Ron began to think that some of the non-physical aspects of our universe, which Dan had pointed out really do exist, just might be traces of "broken spirit." This idea linked FAM with the NSP Paradigm. It was at that point that we realized that the new

NSP Paradigm was the framework that we needed to better understand and explain FAM to others.

But first we had to answer the question: "What exactly did the angels lose when they fell?" Ron said that although the angels losing energy makes sense to us, we were going to get this question. We needed some accepted theology to better describe what the fallen angels lost and what may have fragmented into the energy of our early universe. Joe agreed and said we should ask "The Angelic Doctor" what he thought. The Angelic Doctor is a nickname that had been given to St. Thomas Aquinas a long time ago. He's been called that for many reasons, one of which is for his writings on angels. He wrote more on angels than any other Doctor of the Church. His famous work, the five-volume *Summa Theologica,* has a *Treatise on the Angels* that is about 65 pages long.

The following paragraphs concerning the fallen angels are taken from *The Fallen Angel Model: Deeper into the Mysteries.*

St. Thomas Aquinas, who is commonly called the Angelic Doctor, also referred to them as "darkened." In fact, he had much to say about the fallen angels and what they lost when they sinned and were thrown out of heaven. In his great work, the *Summa Theologica,* he presented his thoughts in a question-and-answer format. Each general topic or question is addressed by several articles, in which there are a number of specific questions and answers. For example, Vol I, Question 64 is: The Punishment of the Demons and the First Article is: Whether the Demons' Intellect is Darkened by Privation of the Knowledge of All Truth? We'll refer to articles by the shorthand notation provided in the Summa. In this case it would be Vol I, Q. 64, A. 1.

The important thing for us to do now is to, as best as we can, consider what the angels lost when they were darkened and thrown out of heaven. In the following few paragraphs, we'll summarize what St. Thomas wrote about this in the *Summa*.

It's important to note that St. Thomas felt strongly that God did not take away anything from the fallen angels that was part of their nature.

"For it follows from the very nature of the angel, who, according to his nature, is an intellect or mind: since on account of the simplicity of his substance, nothing can be withdrawn from his nature, so as to punish him by subtracting from his natural powers, as a man is punished by being deprived of a hand or foot or something else." (Q. 64, A. 1)

He goes on to say in this article that although God did not take away any knowledge that came with their nature, He did take away some of the other knowledge that He had freely given them. That freely given knowledge was twofold: knowledge about "Divine secrets" and … knowledge which "produces love of God." The first kind was not totally removed but was reduced. The second kind was completely removed, resulting also in a complete loss of charity.

In Q. 62, A. 3 and Q. 109, A. 1 St. Thomas says that he believes that all the angels were created in sanctifying grace, which is not something that they would have had by their very nature. However, those that fell lost their sanctifying grace.

In Q. 62, A. 6, St. Thomas argues that more gifts of grace and glory were given to the angels who had more natural gifts.

In Q. 63, A. 7, St. Thomas argues that since the sin of the angels was a sin of pride, and not a propensity to sin, the higher angels were more likely to sin.

In Q. 109, A. 4, St. Thomas argues that the angels are ordered by their nearness to God and those nearer to God have power over those who are further from Him. Therefore, the good angels have power over the bad angels. The bad angels lost their angelic order and some of their power when they fell.

St. Thomas said that when the fallen angels lost these gifts, they lost nothing that was theirs by virtue of their nature. In summary, St. Thomas described what they lost: some knowledge that was given to them, their love of God, sanctifying grace, their angelic order—or standing in relationship to God, and the power that came with that order.

Our understanding of these teachings from St. Thomas is that the increased knowledge and power given to the angels were a result of the "graces" given to them. We think these graces are spiritual entities or spiritual energies. A flashlight is an example in the physical world of a gift that allows a person to see things that by their nature they cannot see. Another example is a car which increases a person's power to move around much better than walking or running. These gifts are physical entities. They don't change a person's nature but give them an ability to know more, see more, and do more. They are entities that can be taken away. We see the gifts of grace as non-physical, "spiritual entities" that when given to the angels increase their ability to gain knowledge and increase their power.

Another, more simple way to think about what the angels lost when they fell is what happens when an electron changes from a higher energy orbit to a lower energy orbit. When this happens, the lost energy has to go somewhere. In this case it is emitted in the form of a photon. According to Christian tradition and theology the fallen angels were created in an enlightened state with God but were "darkened" when they sinned. It's reasonable to think that they lost some kind of spiritual energy. Imagine the amount of energy that must have been released when a large number of angels were darkened and thrown out of the presence of God!

The information from St. Thomas and this line of thinking puts FAM on more solid theological grounds regarding what the angels lost when they fell. We needed this to present FAM to others. *But it is important to point out that the basic concept of FAM is that the fall of some of the angels released some kind spiritual or supernatural energy that devolved, evolved, and became our physical universe.* Whether it was knowledge, power, some consciousness, love, or even some kind of "supernatural infrastructure of heaven" is a matter for future theological speculation and not necessary to pursue any further in this book.

At this point there were still two serious issues remaining that we needed to resolve before we could adequately explain FAM to others. The first is how could something spiritual become something physical and the second is how could a non-physical, spiritual soul interact with a physical body. Both required the Not-So-Physical (NSP) Paradigm because FAM does not make sense in either of the two current paradigms. Obviously, in the materialistic paradigm, there are no angels, and everything in our universe is purely physical. As stated before, we believe this paradigm is inconsistent with the findings of modern science. The other popular paradigm allows for spiritual entities, which are non-physical. However, in this paradigm the spiritual and the material are decidedly different realms. There is no explanation of how entities from one realm can interact with entities from the other realm and no idea that entities could transition from one realm to the other. Furthermore, this paradigm has no concept of "not-so-physical" entities. We believe that this paradigm is also inadequate to explain the results of modern science. Therefore, we believe the NSP Paradigm is not only the best way to interpret the findings of modern science but also is needed to resolve the two remaining issues related to FAM. We believe both of these are examples of what Brother Guy and Father Mueller meant when they said that in some cases a new interpretation of scientific findings is needed to resolve some faith-science issues.

The first of the two issues involved trying to deal with the difference between what we meant by spiritual and what we meant by non-physical, and how could something spiritual become something physical. Here's our best understanding of this issue. All spiritual entities are non-physical. This means, by our definition, that they are not-detectable, not-measurable, and behave in ways that cannot be described by the laws of physics. But we go on to say that there are other non-physical entities that are not spiritual, but they really do exist, and they are in our universe, in our homes, in our physics laboratories, in our bodies, and postulated in physics theories.

So, the word spiritual means more than non-physical: it means somehow involved with God. "He is before all things, and in him all things hold together." — USCCB Bible, Colossians 1:17. We believe that the spiritual energies lost by the fallen angels were suddenly no longer with God. We believe the energies were expelled from heaven along with the fallen angels because they were attached to those angels.

> Then war broke out in heaven; Michael and his angels battled against the dragon. The dragon and its angels fought back, but they did not prevail and there was no longer any place for them in heaven. — USCCB Bible, Revelations 12:7-8.

We think of these spiritual energies as non-physical fields of energy, no longer with God. We believe the loss of being with God made them become unstable and start to devolve and evolve. They didn't enter our spacetime universe; they changed state and became our spacetime universe.

Now let's take a quick look at what physicists tell us about the nature of our universe at its very beginning. They can do this by using Einstein's equations from General Relativity and running time backwards to the beginning. When they do this, it leads to a point in time about 13.8 billion years where the whole universe appears to be a tiny point, just before the equations no longer hold. It's kind of like trying to divide by zero. At this point, there is no matter and even the laws of physics are not fully formed. There is almost perfect order or near zero entropy. Entropy is a measure of disorder used by physicists. Zero entropy means perfect order. Perfect order makes sense if everything in existence just came from a stable, spiritual realm.

Certainly, by any reasonable person's standards, the universe was not physical at this point. From a scientific point of view, it appears to have come from nothing. This description certainly fits well with the idea that spiritual entities, which are non-physical and undetectable, somehow had just become unstable.

The popular physicist Michio Kaku described the very early universe like this:[3]

---

[3] Available at https://youtu.be/RUlVFzl_BJs

...think of a beautiful crystal that shatters...but at the beginning of time when the universe was first created, that's when the crystal existed in its perfect form. We call it the Superforce. A single Superforce held this crystal together. But then we had the Big Bang which shattered this crystal giving us the shattered universe of today. When you look around you, and you see the different forces, mountains, clouds, planets — it's broken. We live in a horribly broken world, but at the instant of creation there was perfection...

This quote comes just after the 4-minute mark in the video. Dr. Kaku also provides a similar discussion in his book: *Parallel Worlds: A Journey Through Creation, Higher Dimensions and the Future of the Cosmos,* on page 84.

Then there was evolution. Some of the non-physical entities changed to not-so-physical entities and finally formed into physical entities like the earth and the sun. So, when the spiritual energy was expelled from heaven, it became non-physical fields of unstable energy. The transition from spiritual to physical goes from spiritual and stable with God, to unstable and devolving/evolving non-physical, then some of it transitions to not-so-physical, and finally some to physical. The description of the evolution of our universe from its non-physical beginning is exactly as described by science.

Physicists cannot determine where the universe came from because the equations break down as we approach the point of creation, and they can't do any experiments to simulate what might have happened before that point.

Note that FAM, without mentioning angels, offers the idea that our universe came from the "breaking," perhaps some kind of symmetry breaking, of what were stable fields of non-physical energy. This is an idea physicists might be able to use whether they believe in angels or not. Right now, physicists are working with what we see as a sort of bottom-up approach. Maybe someday a physicist will postulate examples of non-physical energy fields, that when broken, produce results that are consistent with Einstein's equations describing the very early universe.

On the theological side, it's clear that since God didn't intend this to happen, He could have ignored the lost energy which was forced out from His realm. We believe that He knew some of the angels would sin and He knew there would be consequences. However, like in the one lost sheep parable, God chose to go after the lost energy and guide its evolution to produce as much consciousness as possible. With this idea, God gets the best possible result from the sin of the angels. He produces self-conscious creatures out of the fragmented spiritual energy, which was not conscious when the observable universe began. We believe that when God felt the time was right, He sent His Son to redeem us from this fallen state. Note that it's reasonable to expect that if God created a large number of angels with free will that some of them would fall and that their fall would have important consequences. However, for many people, the fall of both of God's first humans bringing evil on *all* future humans, doesn't seem reasonable and has always been a mystery.

So, this is how we use the Not-So-Physical (NSP) Paradigm to explain the difference between the spiritual and the non-physical and how something spiritual could become something physical.

The NSP Paradigm is also needed to resolve the second issue, which is the final remaining issue. It provides us with a logical possibility for how a spiritual, non-physical soul could interact with a physical body. The body is composed of molecules and atoms, which are made of electrons, protons, and other fundamental particles. These particles are in a physical state when being detected and measured, but for most of the time they are in a not-so-physical state. For example, electrons in an atom "orbit" around the nucleus. However, the "orbit" is not like the earth going around the sun. It's more like a cloud of energy as described in ProWave. The point is that our "physical" bodies are full of not-so-physical entities. Remember when we said we couldn't imagine how a spiritual, purely non-physical soul could interact with a purely physical body? Well, according to the NSP Paradigm, the human body is not purely physical. It contains a lot of entities that are not-so-physical. All of these not-so-physical entities have aspects and behaviors that are physical and non-physical. We believe that a spiritual, non-physical soul can interact with the not-so-physical entities in the human body, which can then interact with the physical entities in the body. In other words, the not-so-physical can act as an interface between the non-physical and the physical because the not-so-physical has both non-physical and physical aspects. Of course, this is a philosophical answer, not a scientific answer, but it is logical. We haven't been able to find any other answer to this question.

The bottom line is that this new NSP Paradigm gave us the ability to take the final steps needed to complete our journey. Taking these steps allowed us to have a reasonable explanation for how God is involved in evolution in a way consistent with traditional Christianity and modern science and evolution.

Once we reached that point, we wrote and published our first book, *The Fallen Angel Model: Deeper into the Mysteries.* It is a treatise on the Not-So-Physical (NSP) Paradigm and the Fallen Angel Model (FAM). That book is intended for those who are interested in the technical details of the underlying scientific and theological issues.

After publication of that book, we began to fully realize just how unusual our 50-year journey had been. We figured that a number of people would be interested in how we got these ideas and in a more non-technical presentation of them. That is, of course, what this book covers.

Our short, non-technical explanation for how God is involved in evolution is provided in the next and final chapter.

Chapter 10

Our Explanation:
How God is Involved in Evolution

We realized many years ago that each of the explanations "on the table" regarding God's involvement in evolution could be placed into one of four broad categories. That is still true today. The categories are *Materialism; Creationism; God Driven, Natural Evolution;* and *God Driven, Natural Evolution with Some Physical Intervention.* We encountered several questions and none of the "on the table" explanations could resolve all the questions to our satisfaction. Our explanation, which is summarized in this chapter, requires a new paradigm, or way of viewing reality, and does not fit into any of these categories. At first, it might appear to belong in the *God Driven, Natural Evolution* category, but it is substantially different from the other explanations and does not belong in this category. With our explanation, what we see in evolution is not God's creative process, but rather is God's salvation process. Our explanation is based on an original sin event and is much more consistent with traditional Christianity than all the other explanations in this category.

The Fallen Angel Model (FAM) is our explanation for how God is involved in evolution. It is a model that makes connections with an event in science and an event in Scripture and theology. As such, it is an interdisciplinary approach that is based on the belief that truth has to be the same in both disciplines but is viewed from different perspectives. With FAM, God is all-good, and God is Spirit. God created angels, which are spiritual, non-physical, self-conscious spirits with free will. Some of them sinned out of pure pride. The fallen angels were thrown out of heaven and that resulted in some lost spiritual energy, perhaps in the form of spiritual gifts which had given them additional knowledge and power. At least some of the lost spiritual energy became unstable and started to devolve and evolve into what we see as our spacetime, observable universe. God chose to work with this disordered, evolving energy in a way that we see as consistent with science to eventually produce life, conscious creatures, and humans which are self-conscious.

However, FAM can only be understood within the Not-So-Physical (NSP) Paradigm which allows for the possibility of some spiritual, non-physical entities to become physical and vice-versa. FAM obviously is not possible inside the paradigm that says only the physical is real. But it also is not possible in the paradigm that says reality consists of separate spiritual and physical realms.

We also believe that the NSP Paradigm is the only way to get a logical interpretation of the quantum reality as discovered and explained by modern science.

In conclusion, we believe that FAM and the NSP Paradigm are the only way to get a reasonable answer to the question: "How is

God involved in evolution?" that is consistent with traditional Christianity and evolution as discovered by science. We further believe that this consistency is necessary for traditional Christianity to survive and thrive into the future.

Although we addressed this question mostly from a Christian perspective, we also believe that Judaism and many other religions should consider FAM and NSP as a way to understand how God could be involved in evolution. Many religions are under pressure these days and need to find ways to remain relevant.

Of course, this kind of journey is never really finished. There are always more questions, more issues, and deeper insights to be found. However, our focus at this point is to share our journey and our insights with others.

We hope you enjoyed coming along with us on this journey and letting us share the process of how we discovered these ideas. We feel like this story is the perfect complement to our more in depth work *The Fallen Angel Model: Deeper into the Mysteries,* published in 2021 by En Route Books and Media, LLC.

www.ingramcontent.com/pod-product-compliance
Lightning Source LLC
LaVergne TN
LVHW051418080426
835508LV00022B/3150